An Introduction to Bee-houses

David F Bates

Northern Bee Books

An Introduction to Bee-houses
© David F. Bates

ISBN 978-1-914934-46-9

Published by Northern Bee Books, 2022
Scout Bottom Farm
Mytholmroyd
Hebden Bridge
HX7 5JS (UK)

Bibliography
"The House Apiary" by John Spiller

Contents

Foreword

This book is intended to be a simple, thought provoking and hopefully enjoyable introduction to the keeping of bees in bee-houses. It is not intended to be a text book on keeping bees since there are many excellent reference books already in circulation.

Whilst the bee-house has existed for many years in European countries such as Switzerland, Austria and Germany it has failed to capture the imagination of beekeepers in this country and currently we have only a limited number of practitioners to-date.

Without wanting to pre-empt the main text of this book I believe that because we have relatively mild winters in this country and assume that our bees can cope easily with British winters we feel we have no need of extra protection and in acting on this assumption we ignore or fail to appreciate the many additional advantages to be derived from operating a bee-house.

I am told it took fifty years for British beekeepers to move over from skeps to fixed frame hives so maybe I should not expect a rush to change. However I do believe that we have allowed others to enjoy this system of keeping bees for long enough and its time we understood and enjoyed the benefits of this approach.

As a past chairman of my local division of the BBKA it makes me sad to see beekeepers having to give up their bees because they lack mobility or have difficulty lifting. Equally we often meet disabled people who would love to keep bees but feel they cannot cope with conventional hives in the open and on rough ground and do not want keep asking other beekeepers for help. Whilst a bee-house would not solve all their problems it would provide a more friendly environment in which to experiment with potential solutions and they may well find they attract other beekeepers who want practical experience of this approach.

The History of Bee-houses

Imagine what it must have been like centuries ago not to have any sweetener other than honey. Most of our food has sugar or some other sweetener added whether it be our breakfast cereal, biscuits, drinks, desserts - the list is endless.

Consider also the many types of wax you use, polish, candles, waterproofing and lubricating agents - then you would only have had bees wax or animal fats.

Only after you have thought for a while about the limited alternatives that were available and the catastrophic effect of being deprived of them can you begin to understand the value of bee products and the need to protect and maximise the production in those days.

Now imagine that you lived in Sweden or Switzerland where temperatures in winter can be -30 degrees Centigrade. I have experienced such temperatures and it is disconcerting to be warned not to go for a walk because by the time you begin to feel cold it is too late to get back! Wild colonies would only survive if they had chosen the right site and retained their full honey production from the previous year - in short natural selection would have been amply demonstrated with productive bees who chose the right home surviving and the rest dying out in the winter or early spring. In addition there was the constant threat from badgers, bears and any other animal who thought the end justified the effort - not to mention the constant threat from humans.

When skep beekeeping was the order of the day production would have depended on how early the beekeeper managed to find a swarm and how kind the weather was to him and his bees. When multiple skeps and fixed frame hives were first introduced beekeepers began to move away from collecting swarms in the spring and destroying the colony in autumn. The quest for the best way of preserving the colony over the winter began. It is not therefore difficult to understand why beekeepers in Europe sort the best way of protecting the colonies they had in order to have an advantage over those relying on spring swarms - the natural human endeavour to help nature in order to help themselves.

There are reports of Swedish beekeepers burying hives in holes in the ground to give protection and of other Europeans placing the hives in cellars during winter. I have seen photographs of hives kept in cages under the eves of barns both for protection from the elements and from animal and human predators. In Britain there are many examples of bee boles - alcoves built into south facing walls to give some degree of protection from wind and weather.

Initial attempts at wrapping up the hives to retain the heat in a hive quickly revealed another problem. They found that the build up in humidity added to the bees' problems and for sometime the demise of a colony was probably blamed on the lack of feeding, a hard winter or disease when in fact the cause was humidity. Some years ago I experienced this problem with a Langstroth hive where the top ventilation holes were blocked with spiders webs and I had reduced the entrance to one bee space. Come the spring the inside of the hive and roof were black with mildew and the outer frames were in a sorry state. The bees were lethargic and took some time to recover. The lesson then is that bees benefit from protection from wind, frost and rain but must receive adequate ventilation.

In Britain, at the beginning of this century, those with money and land were beginning to take an interest in beekeeping partly because more was being written about bees and partly because the benefits of pollination were being realised. Consequently as a result of travelling abroad and seeing bee-houses in Europe they were returning home and casting an eye about their own premises for suitable buildings to be converted. At that time many redundant stables, pigeon houses and outbuildings were converted to bee-houses using conventional outdoor hives. No doubt other country people would have liked to have copied these practices but lacked the premises, time and money to do so.

In Britain the interest in bee-houses never really got established partly because of the lack of facilities and money and partly because beekeepers failed to understand the advantages to be gained. British winters were not that hard and most colonies survived and so practices remained unchanged. What they failed to realise was that apart from winter survival the advantages of reduced winter feeding, quicker start-up in spring, higher honey yields, less weather dependency, simple convenience and calmer bees would more than repay the initial costs involved.

Bee-houses Today

The purpose of the bee-house is to provide an efficient, secure and comfortable environment for the bees and beekeeper whether the bee-house be a small converted shed at the bottom of your garden or a research centre such as those in Switzerland or Germany.

In Britain bee-houses tend to be converted garden sheds or workshops but there is no reason why hay lofts, pigeon lofts, stone out-buildings or even disused chicken houses could not be utilised and this is covered in Chapter 5 on design considerations.

In general, whether in this country or the rest of Europe, ten colonies on one site is considered the upper reasonable limit. More colonies can be kept but unless an adequate supply of nectar and pollen are available additional supplies may have to be provided.

We have seen from the last chapter that bee-houses came about partly due to the need for security and partly due to the advantage of entering spring with a healthy, strong and rapidly expanding colony capable of maximising the yield in the forthcoming season. What started as a necessity in areas with winter temperatures of -20.C for weeks at a time and apiaries at the mercy of badgers, bears and humans soon became a desirable commodity simply because of the additional advantages that a bee-house gave. The problem in Britain has been that because our winters are not severe and losses have been at an acceptable level we have not felt able to justify a bee-house. In recent years however we have started to recognise the advantages of additional protection, convenience and efficiency.

Bee-houses provide protection from the elements. It may be more accurate to say they provide protection from the extremes of the weather since that is the secret of the bee-house. It is not the intention to maintain a high temperature in the bee-house during the winter months otherwise the bees would not hibernate properly and would use up their stores and starve due to excessive activity. The best bees go into a hibernation ball quickly and stay that way for most of the winter using the bulk of their stores to assist in an early and sustained spring build-up. Because of the protection afforded by the bee-house winter feeding can be reduced and because less food is required for maintenance you can expect more honey going into the supers. Frames not covered by bees in winter keep dry and do not suffer from mildew which can be a problem with outside hives.

In his excellent book "The House Apiary" John Spiller stressed that it is not unusual to find that the temperature in a bee house is 10.C degrees higher than the ambient temperature which is due to the heat given off by the hives and would be lost to the air with an outside hive. Due to the heat efficiency bees store more honey, make more wax and make better sections. The temperature inside an outdoor

A bee-house in Somerset using National hives

Fig 1

hive near the wall is only slightly higher than the air outside and it follows that with a bee-house the air temperature in a hive is more evenly spread and easier to maintain thereby releasing bees for foraging and giving those who stay at home better working conditions. In summer the well ventilated and correctly positioned bee-house would provide protection from overheating and reduce the need for water carrying and excessive fanning. Bees are also less effected by cool summer nights.

Whilst there is the additional cost of maintaining the house the need to maintain the hives is greatly reduced since they are not subjected to the full effects of wind, rain, frost and sun. It is far easier to maintain a bee-house than bee hives where the bees have to be transferred to another hive before the work can be done.

The greatest convenience must be that all the equipment, records, etc., can be stored in the bee-house where you need them and never again will you have to dash back to the house to get the item you forgot! In addition you can open a hive without the inconvenience of bees from the other hives getting interested because they will not be aware that another hive is open. It is the best cure for robbing bees whilst opening hives but the important thing is to close one hive before opening another.

For the beekeeper interested in studying his bees a bee-house offers the advantage of being able to watch your

A bee-house of the open type in Holland

Fig 2

bees in comfort and at close quarters, even seated by the hive without undue stress being caused to the bees since they are unlikely to notice you in subdued light unless you move between the bees and the light source. You will become more closely associated with your bees and gain a better knowledge of the condition of the stocks. There is great pleasure to be derived from sitting quietly in your bee-house planning your program for the next few weeks. It is also far easier to set up temperature, humidity and weight recording equipment when you don't have to allow for the weather and hive records can be kept in full view on the hive.

When monitoring the weight of hives the platform type weighing machine should be used and it is important to record the weight at the same time each day. Weight variations of up to 2lbs may be experienced from evening to morning when Lime or Rape nectar is coming in due to rapid dehydration. Also the weight of the hive will vary considerably at different times of the day due to the number of bees in the hive. In winter a hive in a bee-house loses about 1lb per day until January when increased brood begins to add to the weight. Outside hives will lose much more.

A major advantage to the weekend beekeeper and the breeder of queens is that you are never dependent on the weather in a bee-house so plans and time-tables can be adhered to without disruption. You will also notice that the protection and tranquillity of the bee-house seems to calm the bees and reduce their stress levels. Nucleus hives can be stood at a convenient height without fear of them being blown over. Bee-houses lend themselves to the disabled beekeeper. Conventional hives do not need heavy roofs in a bee-house since a cover board is sufficient and if box or drawer hives are used there is little to lift once the hives are in position.

By positioning shelves or tables close to hand you are not lost for somewhere to put things when you have difficulty reaching the floor. The floor is normally boarded or solid so there is no difficult terrain to traverse and you have protection from the weather.

Obviously there are disadvantages but fortunately not that many.There is the initial capital outlay but that must be balanced by the lower hive maintenance, reduced feeding required, faster build up in spring, higher honey yields, convenience, etc. There is a greater risk of disease spread from hives in close proximity particularly with a variety of bee that is prone to drifting such as the New Zealand. The argument of less mobility is a debatable one. Whilst one type of hive may be heavier than another the difficulty of moving hives from a bee-house to an out-apiary are in my opinion no different than moving hives from one out-apiary to another.

Hives and Bee-houses

There are three types of hive which can be operated in a bee-house:-

Conventional single wall hives - National, Langstroth, etc. (See Fig 3)

Box hives. (See Fig 4)

Drawer hives. (See Fig 5)

Conventional single wall hives. There are already many bee-house owners operating traditional single wall hives and doing so quite successfully. It is normal to place them on a low bench or pallets so as to reduce the amount of bending involved and to overcome the tendency to lean over the hive thereby placing your body between the bees and the light source. The bench should not be too high since one needs height above the hive to lift off supers and feed etc. There is no great reduction in the amount of lifting involved and it is still necessary to lift off supers to get at the brood. It is not necessary to have roofs on the hives since a ventilated cover board is quite adequate. It is difficult with this type of hive to make the house bee proof and you may feel there is no real need however you may not appreciate bees being present in the house on all occasions. Concerning the access to outside, in most houses I have seen using conventional hives a simply opening has been made in the wall and the hive is positioned so that the entrance is either against or within a few inches of the opening. There is no reason why a tunnel type connection should not be provided between

National hives in a bee-house

Fig 3

the hive and opening but unless reasonable thought is given to this it can be more trouble than it is worth.

Box hives These hives have been used in Europe for most of this century and were designed for use in bee-houses. The sides of the hives are flat so that hives can be stacked two high and side by side without gaps to stop bees and wasps entering the house around the hives. Normally these hives use conventional frames but with the lugs removed and certainly in Europe brood frames are used throughout both for brood and honey. Access to the brood area is direct since the back of the hive drops down leaving the operator the choice of going into the brood or honey areas. The brood frames are held on a cradle which can be slid out onto the drop down door for inspection. Since these hives were designed for bee-houses they offer a lot of advantages over conventional hives but have some draw backs depending on your style of bee keeping and your bees. Bees that produce a lot of propolis, i.e. in excess of normal bees in Britain, are not best suited to these hives however you may feel that such bees are not best suited to any hive. Also, since these hives tend to have only one honey/super chamber, even if brood frames, they may not appeal to beekeepers who like to leave supers on all year and extract in August. These are hives which operate well where the beekeeper is in attendance regularly and, since they normally have glazed inner doors, allow a cursory inspection of the bees without giving them access to the bee-house.

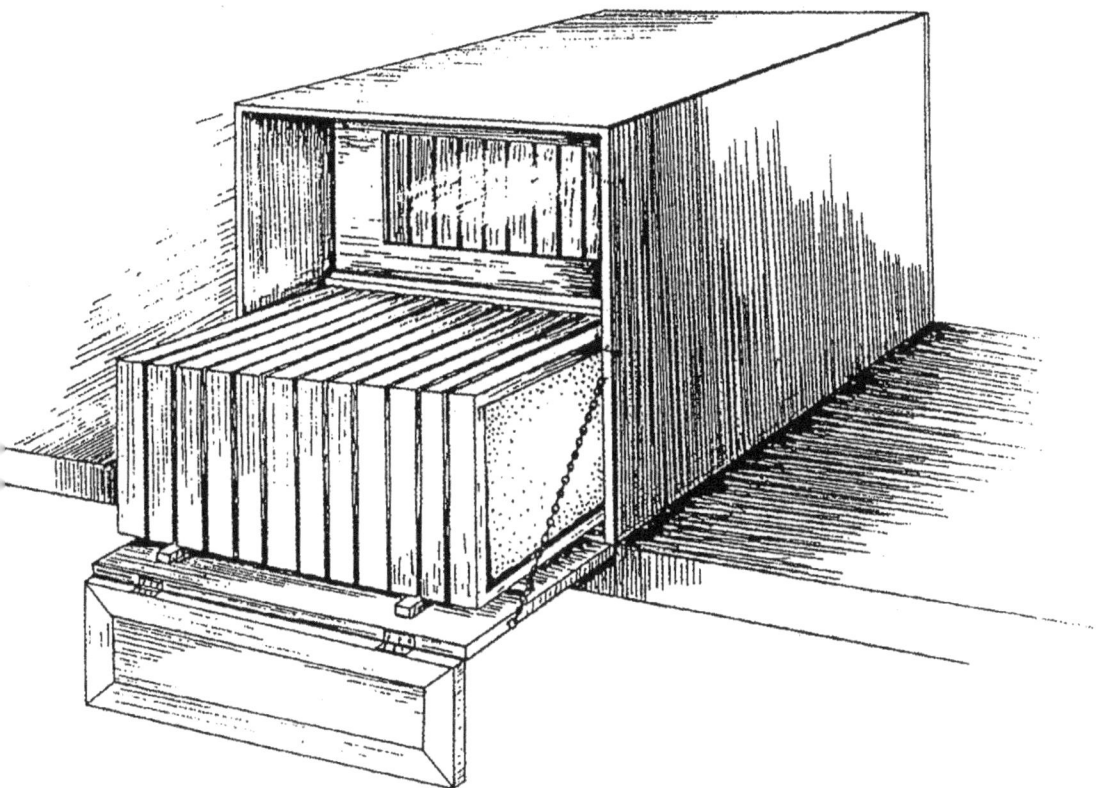

A box type "Euro" hive
Fig 4

Drawer hives Similar in many ways to box hives these operate rather like a filing cabinet where the various chambers can be drawn out for inspection and manipulation. Once in position the amount of lifting is greatly reduced but not a hive to be taken to the rape or beans. Whilst I have no practical experience of them I am told by those that have that they perform very well but must be operated with Carniolan or similar low propolis bees.

Obviously the choice of hive is a very serious one and no doubt many have opted for conventional hives simply because that is what they already had or they indulge in migratory beekeeping and want to use one style of hive throughout. For those who wish only to operate a bee-house and have no plans to move hives to rape or beans then there is a good case for box or drawer hives which are purpose made for bee-houses. Whilst box and drawer hives are not designed to be used outside without a roof or protection from the elements many European beekeepers place them in covered trailers and do move them from crop to crop. You really have to consider what you want out of beekeeping and your hives and if its only fun then there is no reason why you cannot operate both as I do. I trust that has confused you!

If you are considering a bee-house to combat a bad back or a disability then you must be aware that conventional hives, even when placed on a stand to reduce bending, will entail a lot more lifting than a box or drawer hive. In addition, because the hive is raised the third super maybe at shoulder

A Drawer type hive
Fig 5

height and whilst it may be easy or at least manageable to lift a full super from waist height to a bench it may be very different trying to lift a 25lb super from bench to shoulder height. If you have any doubts carry out a few trials with lesser weights to make sure you are happy with what you are planning and that you will still be happy in five years time.

We all know that bees do not like high humidity but I believe they hate permanent dampness even more. I have seen hives in Holland with perspex cover boards and no top ventilation but with varroa screens left in place to give a bottom through ventilation and the bees looked very happy. I have also seen box hives in a shed with a permanent leaking roof which kept the hives damp at least and wet at worst and no amount of ventilation will overcome that, the bees looked very sad by Spring but picked up when put outside with a temporary roof. The bee-house is there to protect the bees and hives from the worst of the weather not to create a detrimental enviroment.

When talking about protection from the weather it is easy to ignore the problems of Summer and you would be well advised to remember that honey will spoil if held above 65 degrees centigrade for any time, wax will melt or start to collapse at 60 degrees and brood will die above 38 degrees. Wooden buildings covered in black roofing felt can heat up very quickly and whilst the subject of ventilation is covered in the chapter on design it is worth mentioning this often ignored problem. I would strongly recomend the use of a max-min thermometer to record the temperature variation throughout the day. You may be surprised to discover that the temperature peaks at 3.00pm but is always ok at 5.00pm when you get home from work.

Bees and Bee-houses

It is not the purpose of this chapter to go into great detail about the way bee varieties have developed but rather to attempt to identify the strengths and weaknesses of each and relate that to their suitability for use in bee-houses.

Broadly speaking the darker bees have their origins in cooler areas of Europe and the yellow bees from the warmer areas of the Mediterranean countries. More important perhaps is that their habits relate to the climate and seasons of those areas.

In a perfect world we would have bees that gathered immense amounts of honey, were not aggressive to their handler but defended well against robbing bees, produced a minimum amount of propolis but made beautiful white cappings, used a minimum of winter stores but built up quickly in the early Spring - and of course never swarmed and didn't have a sting! Now look at the "average" bee in Britain and you will have difficulty finding many of the virtues mentioned above.

The original British bee disappeared due both to the "Isle of Wight Disease" in the first years of this century and the importation of foreign bees starting probably with the importation of Ligurian or Italian bees in 1859. Consequently most bees in this country are now simply mongrels and some would say not very good ones at that. This then raises a problem when it comes to suggesting a suitable bee for use in a bee-house or indeed any system of management since the problem becomes not what is best but where can you get the pure bees that are best and how do you keep them pure. Another concern should also be what happens if they cross with a "local" bee. Are you going to be creating a major problem for local beekeepers and neighbours by producing a bad tempered strain?

The answer is that whilst there are always sources of particular varieties advertised in the bee press the real problem is keeping them pure. That can only be done by instrumental insemination or regular Queen replacement unless we ever get to a point where every drone flying around your area is of the same variety, which will only happen if we have a national and effective bee breeding program. With the introduction of Varroa we probably have the best chance now that we have ever had because wild swarms will be destroyed or reduced in numbers and the number of rogue drones will be reduced. The Germans have had a program for many years and are now recognised as skilled breeders of good bees. I should now stress that they have the best bees for their country but the best bees for Britain would be selectively breed bees from the best that we have in Britain.

Having said that let us look at some of the varieties and highlight their strengths and weaknesses:-

CARNIOLANS

These are extremely gentle bees and have a grey appearance due to the white to grey hairs on their abdomen which are longer than on the European or Italian bees. Their home is Austria and Yugoslavia and consequently because of the hard winters they over-winter well and build up fast in spring. They are ideal for box and drawer type hives in bee-houses since they produce little propolis. Their weak point is their tendency to swarm mainly because they are very prolific and need ample room. The Germans based their bee breeding program on this variety although the comment has been made that with the benefit of hind sight they should have taken their local bee because of the Carniolans tendency to swarm. It is alleged that a cross between a Carniolan and Caucasian is potentially a very vicious bee whereas when crossed with the Italian gives a hardy docile strain, building good sections. Like most dark races they produce fine white cappings. They hold well on the comb and are reputed to be more difficult to shake than most.

CAUCASIAN

These bees were first imported into this country in the 1930's. There are several distinct races in the Caucasus which vary in colour from grey to yellow and in characteristics from Carnolian to Italian. They received a lot of attention because of the length of their tongues, their temper and their hardiness. They are moderate swarmers and are reputed to settle much higher than other swarms which is a nuisance if like me you don't have a head for heights! The bad news is that they produce a lot of propolis which is not good news for those of you who use box or drawer hives in your bee-house. In general they have a reputation for producing vicious crosses and with the number of mongrel and Italian drones flying around Britain must give cause for concern.

ITALIAN/NEW ZEALAND

The Italian bee used to be called the Ligurian bee and although I have grouped it with the New Zealand because they are to me all of the same type I am sure the New Zealand beekeepers will prefer to see their bee as an improved Italian.That said it is a handsome bee varying from yellow to light orange in colour and in general calm and docile and although ill tempered colonies are occasionally found I suspect they are not pure. They have a reputation for robbing and drifting and a beekeeping friend in Somerset assures me there is always at least one yellow bee cruising round his hives looking for a way in.

Since they are basically of Mediterranean origin they have difficulty in gauging the end of a British season and continue to raise brood well into the winter and are slow to build up in spring. This means that they tend to need a lot of feeding and risk starving in early Spring as well as using up honey in June that they collected in May. They prefer dry winter conditions and are thought to need more protection in colder climates.

BRITISH

It is tempting to say that to write at length on this variety is to add strength to the accusation that we live in the past instead of looking forward simply because this bee is extinct to all intents and purposes.

It was a dark brown and not black bee and had many good points in its favour. They produced excellent white and even cappings which were slightly domed and were therefore valued for the sections they produced. They were moderately tempered or to be more accurate they were generally well tempered but ill tempered colonies were not uncommon. Their faults were that they did not defend very well and were poor house keepers and were susceptible to disease. It is therefore small wonder that they were ravaged by "Isle of Wight disease" and that those that survived got a rough time from the Italian bees being imported at that time and hybridised by the rest. Whilst there are rumours occasionally that a colony of pure bees have been found we must except that they are very much in the minority. Perhaps it would be kinder to say that the "British" bee is now being developed from the better bees resulting from fifty to one hundred years of hybridisation in this country but until beekeepers in general truly recognise the advantages and work in full cooperation with those doing this work we will only end up with limited pockets of good "British" bees.

Design Considerations

Perhaps before we get into detail on this subject we should again consider what we are trying to achieve with a bee-house. We would expect a good design to give protection to the bees in terms of temperature, rain, wind, humidity, vibration, disturbance and predators it order that they may prosper better than their cousins outside and make our investment in materials and time worthwhile. For the beekeeper it should be a pleasure to work in a bee-house in terms of a comfortable environment, convenience and safety. It should not make us feel as though we have spent a few hours in a sauna, unable to move freely, with nowhere to put anything down and in the company of a few thousand bees who could not find their way out.

We now come to decision time in the sense that we should give very serious consideration to certain aspects of bee-houses since the effect of these decisions will have a major impact on our future enjoyment and success. A director of a company once gave me the advice that there are two types of decision, those that could be changed with little inconvenience, disruption or cost and would not effect the profitability or performance of the company - eg what colour shall we paint the toilets? - these you make today and change tomorrow if you don't like them. The other type are major decisions and if you get them wrong they will be expensive, disruptive and will have a major effect on your company and you - these you take your time over and make sure that you have as much information and input as you can get.

Size Probably the most important long term decision required and one for which you will need to provide additional information is what size your bee-house is going to be. First you must ask yourself how many hives are you likely to want to keep in the bee-house over the next five years. Be realistic bearing in mind that few areas in the UK can support more than ten hives on one site without additional provision. Nucleus hives can be placed outside the house and only be brought in when a colony is established. What else are you going to do in your bee-house - extract honey, instrumentally inseminate Queens, store equipment, store your apples in the winter! Now is the time to take all these things into consideration.

What type of hive are you going to use? If a conventional single walled hive you don't need a hive roof only a ventilated cover board. You may want the hive to be on a raised bench to reduce lifting, and you will need to be able to lift at least an extra super high to allow for the unforeseen such as lower frames stuck to upper frames, etc. If you decide to use box or drawer hives which are the traditional hives for bee-houses you must allow at least a meter (a yard) in depth for the hive, the drop down door or drawer and space for yourself. This is important because you do not want to end up with a cramped working area which will eventually drive you mad! You should also allow for a strong shelf or bench to put things down on and to give you an area to work on equipment. After allowance for the hive, working space and shelf space you probably won't get away with much less than 8 ft. (See Fig 6) I have purposely switched to Imperial measure for a moment since most sheds come that way ie 6 x 4, 6 x 8 and 8 x 6, etc.

Roof Style There are two main types of roof styles. (Fig 7)

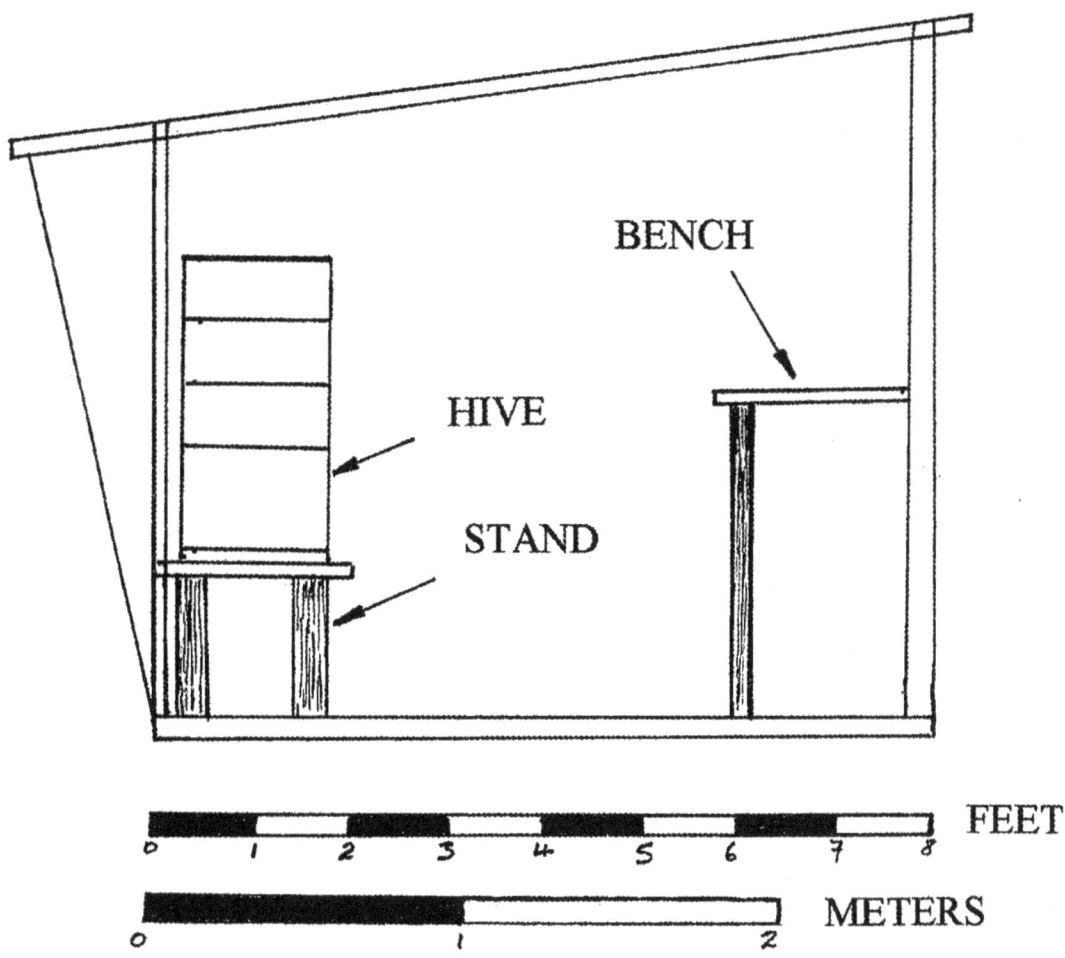

An indication of space required in a Bee-house
Fig 6

The pent roof which is a one slope roof from the high side to the low side with the window in the high side and a door in the end. The other type is the ridged or apex roof which is high in the centre and slopes down to the longer side walls. Whilst you may not have a choice over this you could consider the following. With a pent roof the tendency may be to place the hives against the high sided wall but this leaves you permanently standing where the roof is at its lowest so consider placing the hive on the low side and standing on the high side. The only problem with this approach is that the window is normally on the high side and you would be working between the light and the bees which is not perfect when you consider the section on lighting and bee clearance. With a ridged roof you could consider locating the hives at the end wall to give more working height and space for you. Irrespective of the roof type consider extending the roof and side walls over the hive entrances to protect them from wind and driving rain. (See Fig 8)

Location You can relocate a bee hive simply by taking it three miles away for a few days and then bringing it back again to the new location. If you only want to move the hive less than three feet that is not a problem but if you move the hive ten feet away the bees will congregate were the hive was and not realise it is only ten feet away. This is probably due to them learning the layout of the immediate area around their landing approach but failing to see their hive ten feet away once they have landed. To overcome this problem with a bee house would require taking all the bees away and bringing them back again to the relocated bee-house - not the end of the world but something you could do without.

You have three areas to consider, your family, your

PENT ROOF

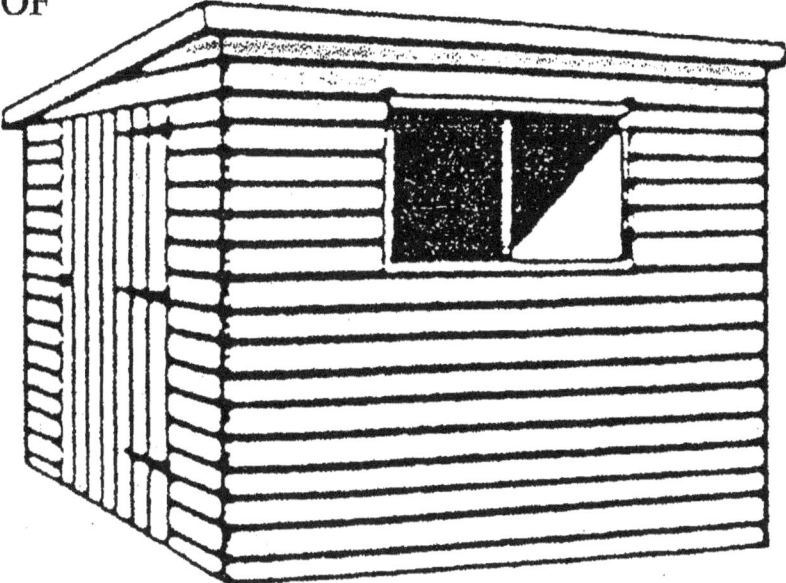

APEX/RIDGE ROOF

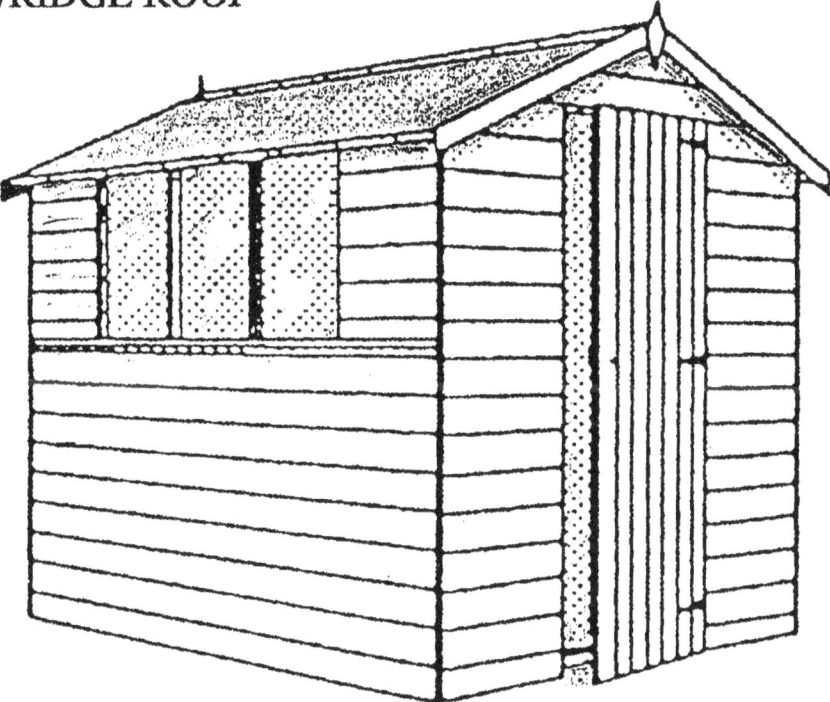

31

neighbours and your bees. If the hives face across your garden the garden could well become a no-go area for your family. Whilst I have heard it said that extracting honey in the kitchen and getting your partner to help you reverse a caravan are major contributors to divorce, putting your garden off limits to the family must be a close third. When bees leave the hive they tend to fly away climbing gently as they get further from the hive. If however you position them a few meters from an obstacle such as a wall or hedge they quickly learn to climb rapidly in flight and continue to do so after clearing the obstacle. My out-side hives face onto a lane at the end of my garden which is bounded by a two meter high hedge, consequently I have never had complaints from passing pedestrians because the bees are at about four meters when they pass over them. Many of my neighbours were not even aware that I keep bees until I put the "Honey for Sale" sign outside. Concerning the direction that hives should face the general consensus of opinion is south or south-west when ever possible. East opens them to cold winds and can encourage them to fly on cold but sunny winter mornings when the front of the hive can be warm but the air temperature is so low they never make it back to the hive. North is both cold and dull which tends to cause late starts and early finishes as far as workers are concerned and generally is not ideal. Even westerly can present problems from the prevailing wind if no wind-break or protection is given. Whilst it is often difficult to get the perfect situation try to aim for a location that gives sun on the front of the hive in the morning and protection from direct winds on the hive entrance - if you can achieve that your bees will not have much to complain about.

Example of an extended roof and side walls
to reduce draught and rain at the hive entrance
Fig 8

A rotating window to aid bee clearance
Fig 9

Lighting and Bee Clearance I link these together because the light source must also be the escape route for bees in the bee-house. Bees will tend to rise to the highest point in the bee-house unless there is a source of light which will normally take priority. Consequently it would be wise to link the two together for maximum efficiency. Any form of bee escape should not be of a permanent nature - it should only be open when you want it to be and at all other times it should be closed since I believe that bees learn and there is the risk that they, and wasps, will start to get into the habit of re-entering the house via the bee escape. In an ideal world I believe the best light source and bee escape is a roof light which can be opened as required, in fact I wonder why I would need a side window at all which from a security point of view is a weak point. If you have a roof light and a side window you will find that many bees will go straight to the side window and stick there. In this situation either cover the window before you start to open hives or have the sort of window which is centre hinged and can be rolled over or around so that the bees suddenly find themselves on the outside. (See Fig 9) Bees exiting via these openings will not attempt to return by them since they haven't learnt of them - they only know of the hive entrance and will return to the hive that way. Another form of light/bee escape is a long shallow window above the hives on the side wall which works quite well because you are not between the bees and the light source. In general life is easier if the light source is above or in front of you. If it behind you they notice every movement you make.

Insect Proof Ventilation A bee-house without ventilation, especially if it is covered in roofing felt will rapidly become an oven in summer. If you put a roof light in as well it can only get worse. The purpose of the bee-house is to help even out the extremes of the weather and in the summer one of its functions will be to shade the bee hives and the beekeeper. On a hot summers day it would be unwise to just leave the door open since this would encourage not only wasps and bees but also spiders and you do not need a house full of webs when your bees are trying to exit the bee-house. It follows therefore that any ventilation should be screened to exclude insects. A single ventilation opening is not very effective and you should aim for through ventilation by having at least two openings on opposite sides of the house. Bear in mind that hot air rises up into the roof area and that when the wind blows at a building it builds a pressure on the side the wind is blowing from but a vacuum on the other side of the building consequently it would seem logical to put the upwind ventilator low down and the down wind ventilator up high - are you confused then consider a screen that can be put in place of the door when it is left open - the possibilities are only limited by your ingenuity. The ventilators do not have to be permanently open and flaps should be provided to close them when required. There could well be value in the automatic greenhouse window openers which could be adapted to suit the ventilators in your bee-house.

Storage Whilst recently in Germany I was taken to a farm where a farmer had converted a hay loft to a bee house some fifty years ago. Along one wall were box hives stacked two high and totalling fifty. Along the other wall were two meters high cupboards looking rather like built-in wardrobes. When I opened these 'wardrobes' I found they were full of racks holding clean drawn out frames. It was late September and all his extracting was finished, frames had been cleaned up by the bees and the bees had been fed and treated for Varroa. It was made more impressive by the fact that he was not expecting us and so there had been no preparation for my visit. The doors of each frame cupboard were sealed with felt strips. I now aim to be this efficient and organised - one day!

Meanwhile my frames are wrapped in newspaper, because wax moths seem not to like the smell of news print, and then placed in cardboard boxes just as they would hang in the hive otherwise there is a risk the wax may sag on a warm autumn or spring afternoon. You must also guard against mice getting access as they love to build a nest in frames.

Concerning items of equipment there are four major areas available to you, on shelves around the bee house, under the bench that the hives rest on - if you use such a bench, under the work table - if you have one, or, above the hives if you use box or drawer type hives. Of these I think I would favour above the hives. Try to think in terms of "if I'm stood holding a frame with one hand could I reach everything I'm likely to need with the other"? This may sound a little over the top but it is a measure of how accessible the every-day items of equipment are and should not be unreasonable in your perfect bee-house.

Final Thoughts A potential problem with a bee-house is vibration caused by the beekeeper walking around on a wooden floor and by loose doors banging shut in the wind. If you have ever tapped the side of a hive in the evening and heard the rise in tone of the bees you will appreciate that the long term effect of regular knocking or vibration can be irritable bees, constantly on their guard and being a little over protective. It will help tremendously if the hive support stand is not resting directly onto a wooden floor but in some way insulated from it. One solution maybe to allow the legs of the hive stand to pass through the floor onto bricks resting on the ground below. Whatever solution you decide on relative to your situation this extra thought and effort will be well rewarded.

Important Tannalised wood is protected both from rotting and insect attack and gives off fumes which will kill bees. A friend had built a new bee-house and shortly after installing the bees he noticed that his bees seemed to be suffering from bee paralysis. There were no dead bees outside but many on the floor of the bee-house. The floor was made from tannalised boards and after sealing the floor with varnish the problem immediately cleared up. Shortly after that, however, he replaced the floor to be on the safe side. I wonder what it does to beekeepers?

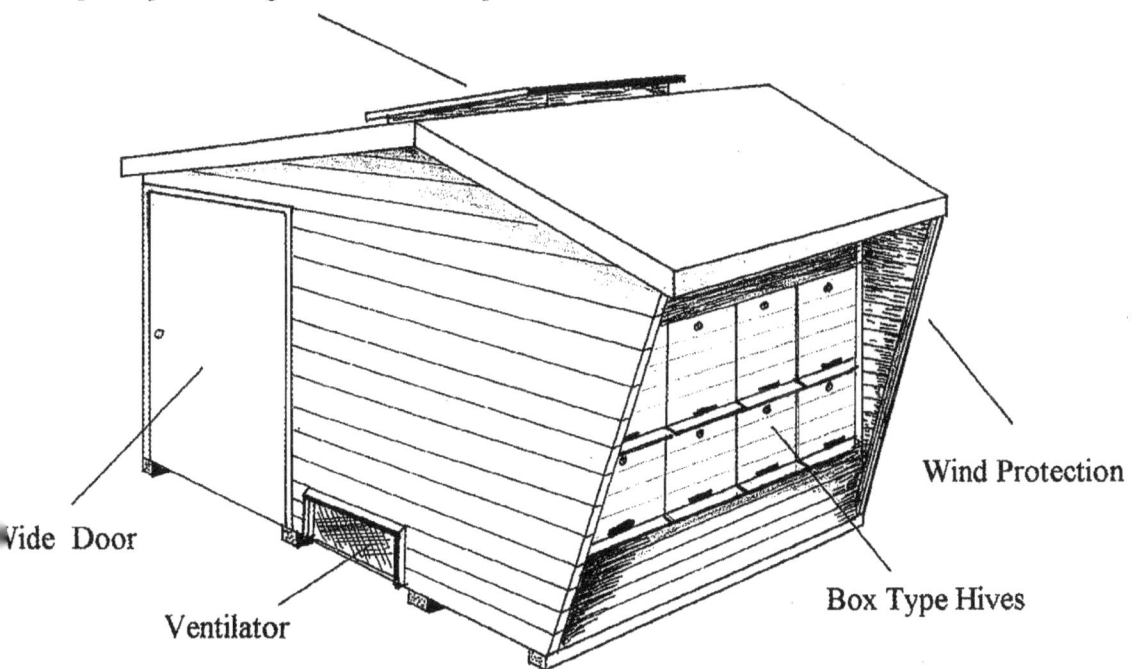

Fig 10

Acknowledgements

Manfred Bender, Taunusstein, Germany : The man who introduced me to bee-houses, fired my enthusiasm and showed me that the manipulation of bees can be elevated from a craft to an art form.

Peter Kinsey, Harrow-on-the-Hill, Middlesex : Who kindly provided the photograph of the drawer hives used in his bee-house where he breeds pure Carniolan queens using artificial insemination.

Phil Fouracres, Wrantage, Somerset : Who has operated a bee-house using National hives for many years and allowed me to take photographs for this publication.

Tony Elcomb, Wincanton, Somerset : Who operates a bee-house using National hives and allowed me to take photographs for this publication.

www.ingramcontent.com/pod-product-compliance
Lightning Source LLC
Chambersburg PA
CBHW040154200326
41521CB00019B/2607